THE N
NYACK, N. Y. 10960

W9-CGT-394

TRICKY
TWISTERS

THE NYACK LIBRARY
NYACK, N. Y. 10960

TRICKY
TWISTERS

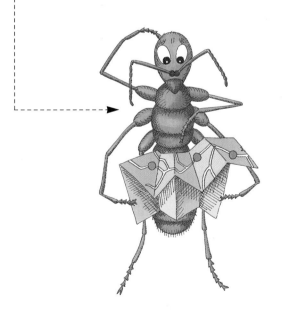

WORLD BOOK, INC.
CHICAGO LONDON SYDNEY TORONTO

© 1997 World Book, Inc. All rights reserved.
This volume may not be reproduced in whole or in part in any form without prior written permission from the publisher.

World Book, Inc.
525 W. Monroe
Chicago, IL 60661
U.S.A.

Editor: Melissa Tucker
Design: Lisa Buckley
Cover design: Design 5

Library of Congress Cataloging-in-Publication Data

Tricky twisters.
       p.   cm. -- (World Book's mind benders)
   Summary: A collection of puzzles, games, and riddles emphasizing logic and deductive reasoning, some of which involve numbers.
   ISBN 0-7166-4109-7 (softcover)
   1. Mathematical recreations--Juvenile literature.   [1. Puzzles.
2. Mathematical recreations.]   I. World Book, Inc.   II. Series.
QA95.T745 1997
793.7'4--dc21   97-6018

For information on other World Book products, call 1-800-255-1750, X2238, or visit us at our Web site at http://www.worldbook.com

Printed in Singapore.

1 2 3 4 5 99 98 97

Introduction

The puzzles you are about to do require a lot
of careful thought—they are brain twisters!
There are lots of clues, but you have to sort
out all the clues. You have to think about
what each clue really means. And you may
have to fit them together in different ways to
find the answer.

It is best to use pencil and paper, so that
you can write down all the clues and keep
them straight in your mind. In some cases, it
will be a big help if you draw diagrams. In
some cases, you can solve the puzzle by just
crossing out clues, until suddenly the right
answer pops out at you!

The museum

Professor Priscilla Pippen was a history teacher at a college. One Saturday, as she drove through a very small town, she saw a tiny building with a sign on it. The sign read, "Museum of Natural History." Professor Pippen parked her car and went into the museum.

There were only five exhibits in the museum. These were:

1. A prehistoric arrowhead made of copper.

2. The fossil skeleton of a dinosaur no bigger than a chicken.

3. An ancient Roman coin marked with the date 120 B.C.

4. A red diamond in a ring.

5. An ancient Egyptian cat mummy.

Professor Pippen knew at once that one of the exhibits was a fake. Which one was it?

(ANSWER ON PAGE 29)

The languages of Kabulistan

On the planet of Kabulistan there are two groups of creatures, the Swazzish and the Buzziks. Each group has its own language. Many of the creatures, however, speak both languages. But 36 per cent cannot speak Buzzik. And 19 per cent cannot speak Swazzish. What per cent of the creatures of Kabulistan speak both languages?

(ANSWER ON PAGE 29)

Busy ants

Many kinds of ants make underground nests. The nests often have many tunnels leading up to the surface. Ant workers scurry in and out of these tunnels. Some go out to look for food. Others come back with food they have found.

Let's say that one ant nest has three tunnels leading up to the surface. Three worker ants are out looking for food. They all come back at the same time. How many different ways can they go into the nest, each using a different tunnel?

(ANSWER ON PAGE 29)

Lost equipment

A group of children went camping. They lost the eight important things shown at right. The eight things are all hidden somewhere in the picture below. Can you find them?

(ANSWERS ON PAGE 29)

tent

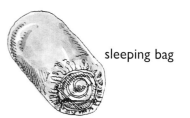

sleeping bag

coffee pot

first-aid kit

knapsack

campfire

flashlight

canoe

A crowd of cows

If you were to see a cow in front of
two cows, a cow behind two cows,
and a cow between two cows, how
many cows would you see altogether?

(ANSWER ON PAGE 29)

Chipper's relative

A raccoon named Chatter saw a friend of
his, named Chipper, talking to a young
raccoon. As Chatter neared them, the
young raccoon trotted off.

"Who was that little raccoon you were
just talking to?" asked Chatter.

"Why, that raccoon's father is my
father's only son," answered Chipper.

In what way were Chipper and the
young raccoon related to one another?

(ANSWER ON PAGE 29)

A trip downtown

Geraldine and her mother went downtown. They rode the bus, because Geraldine's father had taken the car to work. They visited a doctor, cashed a check, and bought a roast for supper.

Geraldine's father doesn't work on Saturday or Sunday. The doctor's office is closed on Wednesday. The bank where the check was cashed is open Monday, Wednesday, Friday, and Saturday. The butcher shop is closed on Friday. What day did Geraldine and her mother go downtown?

(ANSWER ON PAGE 29)

Ginny and Sarah

Sarah is now the same age that Ginny was four years ago. Four years ago, Ginny was twice as old as Sarah. Ginny is now twelve. How old is Sarah?

(ANSWER ON PAGE 29)

Shakes all around

Six children were nominated for the position of president of their school class. Doris won the election. She shook hands with each of the other children. Then, all the children shook hands with each other. How many handshakes were there altogether?

(ANSWER ON PAGE 29)

Mr. Short, Mr. Tall, and Mr. Middle

Three friends, all men, were traveling on an airplane. They struck up a conversation with a lady who was seated near them.

"Believe it or not," said the tall man, "our last names are Tall, Short, and Middle!"

"That's right," chuckled Mr. Short. "And as you can see, one of us is tall, one is short, and one is middle-sized. But not one of us looks like our name."

Can you figure out the name of the short man, the tall man, and the middle-sized man?

(ANSWER ON PAGE 30)

Three puppies

Three puppies were walking one behind the other.
The puppies' names were Spot, Ginger, and Bingo.
Spot was behind Bingo. Ginger was in front of
Spot. Bingo was in front of Ginger. Which puppy
was first in line?

(ANSWER ON PAGE 30)

Who earns the most?

Bill and Ellen had just begun work at their first jobs and were talking about them.

"I get paid $5,000 for the first six months," said Ellen. "Then my salary goes up $50 every six months from then on."

"My salary is $10,000 a year," Bill told her. "I get a $200 raise at the beginning of each year."

If they both keep working at their jobs for five years, who will earn the most money?

(ANSWER ON PAGE 30)

What's the number?

A mother, father, and two children were driving from one city to another for a vacation. The car was moving at a speed of 55 miles (or 55 kilometers) an hour.

Suddenly, the father, who was driving, noticed something. "Look at the dial that shows how far we've come," he said. "It shows the number 11911. That number is the same backward as it is forward. That's unusual. I bet it will be a long time before the dial shows a number like that again!"

But he was wrong. Only two hours later, the dial showed another number that was also the same backward as it was forward. What was the number?

(ANSWER ON PAGE 30)

Squirrels and nuts

Two squirrels named Chatter and Chitter were gathering nuts for winter. They had agreed to divide all the nuts they gathered into two equal shares.

Each squirrel gathered all the nuts he could find. Then the two got together to divide up the nuts. Unfortunately, squirrels aren't very good at arithmetic. It took them three tries to divide the nuts equally!

On the first try, Chatter gave Chitter as many nuts as Chitter already had. But that didn't make things even, so they tried again.

On the second try, Chitter gave back to Chatter as many nuts as Chatter now had. But that still didn't make things even.

On the third try, Chitter gave Chatter 10 more nuts. That made everything right, because now each one had 50 nuts.

How many nuts had each squirrel collected?

(ANSWER ON PAGE 30)

18

The coin test

Long ago, people weighed things in a kind of scale called a balance. A balance is simply two pans, one hanging from each end of a long rod. The rod is balanced across a bar—like a teeter-totter. When an object is placed in each pan, the pan with the heaviest object dips down. If both objects weigh the same, the pans balance.

Once upon a time, in the city of Samarkand, a young girl went to a merchant to ask for a job. The merchant decided to test the young girl's cleverness. She showed the girl nine gold coins and a balance.

"These nine coins look exactly alike," said the merchant. "But one of them is a counterfeit. It isn't pure gold, so it does not weigh as much as the others.

"You could easily find the counterfeit coin by weighing all the coins, two at a time," the merchant continued. "If two coins balanced, you'd know they were pure gold. If one was lighter than another, you'd know it was the counterfeit. But, to do that, you might have to use the balance as many as four times. Can you find the counterfeit coin using the balance only twice?"

How could the young girl do it?

(ANSWER ON PAGE 30)

The five hobbyists

In a small town live five ladies who are close friends. Their names are: Ms. Schmidt, Ms. Rodriguez, Miss Fujimoto, Mrs. Kowalski, and Mrs. Robinson.

Each lady has a different hobby. One is a gardener. One makes pottery. One knits. One is a painter. One writes stories.

From the following six clues, can you tell which lady has which hobby?

1. Mrs. Kowalski doesn't care for gardening.

2. The gardener and Ms. Schmidt went to school together.

3. Mrs. Kowalski and Miss Fujimoto wear scarves that were made by the knitter.

4. Miss Fujimoto and the writer often have lunch with the gardener.

5. The gardener, the knitter, and Ms. Rodriguez have some of the vases made by the pottery maker.

6. Mrs. Kowalski asked Ms. Rodriguez if she thought the painter would do a portrait of her daughter.

(ANSWERS ON PAGE 31)

Frank's folly

Two boys, Frank and Larry, were walking home from a candy store. They had just spent their whole allowances on candy!

Frank sighed. "I wish I had twice as much candy as I have now," he said.

"How much do you have?" asked Larry.

Frank told him. Larry, who was very good at arithmetic, thought for a while. Then he said, "We have to walk three more blocks to get home. So, I'll tell you what—at the end of each block, I'll give you enough of my candy so that you'll have twice as much as at the start of each block. But each time, you have to give me back eight pieces so that I'll have some, too. Is it a deal?"

It seemed like a great idea to Frank. "Sure!" he exclaimed.

At the end of the first block, Larry gave Frank enough candy so that he had twice what he started with. Frank then gave back eight pieces.

At the end of the second block, Larry doubled Frank's candy again. He got back eight pieces.

At the end of the third block, they did this again. But this time, when Frank gave back eight pieces—he had nothing left!

How many pieces of candy did Frank have to start with?

(ANSWER ON PAGE 31)

22

The dull movie

A very dull movie was being shown at the
Cinema Theater one evening. The movie was
so bad that one-third of the audience left
during the first part of it. One-half of the rest
left during the middle part. And one-half of
those left in the theater walked out during the
last part. Only 25 people were in their seats
when the movie ended.

How many people were in the audience
when the movie began?

(ANSWER ON PAGE 31)

A nutty puzzle

Nicodemus Nubbs sold nuts. One day, Nicodemus went to town with a large basket that was full of nuts. He quickly found a customer—a lady bought half the nuts in his basket, plus half a nut!

A short while later, another customer hailed Nicodemus. This man bought half the nuts in the basket, plus half a nut.

Soon after, there was another customer—a little girl. She bought one-third of the nuts Nicodemus had left.

Nicodemus now had exactly one dozen nuts in his basket. All the nuts were whole, and all the nuts he had started with had been whole. Nicodemus hadn't had to cut any nuts in half when he made his sales.

How many nuts did he start with? How many nuts did each customer buy?

(ANSWERS ON PAGE 31)

The lunch bunch

Six children rode their bicycles to McMullin's Hamburger Restaurant for lunch.

1. Matt had what Mary Jo did, except that he did not have an ice-cream sundae.

2. Diana had what Brian had, plus a hamburger.

3. Gail had what Judy had, but without French fries.

4. Mary Jo had what Judy had, but she had pop instead of a milk shake.

5. Brian had the same things Matt did, except that he did not have a hamburger.

What did each child have for lunch?

(ANSWERS ON PAGE 32)

What's missing?

The pictures on these two pages seem to be just the same. But if you look closely, you'll see they're not. Some of the things in the

picture on the left-hand page are missing from the picture on the right-hand page. Can you find what things are missing?

(ANSWERS ON PAGE 32)

A hungry bookworm

Two large books are side by side on a shelf. The first book, on the left, contains two hundred pages. The second book contains one hundred pages.

A hungry bookworm got into the first book. It ate its way through the first page of the first book and all the way through the last page of the second book. How many pages (not counting the covers) did the bookworm eat through?

(ANSWER ON PAGE 32)

Answers

The museum (PAGE 7)

The fake exhibit was number 3—the ancient Roman coin marked 120 B.C. The abbreviation B.C. stands for "Before Christ." People who lived 120 years before Christ couldn't possibly have known it, of course! So they would not have put B.C. on any of their coins.

The languages of Kabulistan (PAGE 8)

We know that 36 per cent of the creatures speak only Swazzish. And 19 per cent speak only Buzzik. That's 55 per cent who speak only one language. This means that 45 per cent of the creatures speak both languages.

Busy ants (PAGE 9)

There are six possible ways. Number the tunnels 1, 2, and 3. Call the ants A, B, and C. Then you can see how this works out:

Ways	Tunnel 1	2	3	Ways	Tunnel 1	2	3
(1)	A	B	C	(4)	C	A	B
(2)	A	C	B	(5)	B	C	A
(3)	B	A	C	(6)	C	B	A

Lost equipment (PAGE 10)

A crowd of cows (PAGE 12)

You would see three cows standing in a line. One would be in front of the other two, one would be behind the other two, and one would be in the middle.

Chipper's relative (PAGE 12)

The young raccoon's father is Chipper's father's only son. If Chipper's father has only one son, that son has to be Chipper. Therefore, Chipper must be the young raccoon's father.

A trip downtown (PAGE 13)

Geraldine's father does not work on Saturday or Sunday. We know he was at work on this day, so Saturday and Sunday are ruled out. The doctor's office is closed on Wednesday, so Wednesday is out. The bank where Geraldine's mother cashed a check is closed on Tuesday and Thursday, so these days are out. The butcher shop is closed on Friday, but Geraldine's mother bought a roast there. So, Friday is ruled out. This leaves only Monday.

Ginny and Sarah (PAGE 13)

Ginny is now twelve. Four years ago, she was eight. If Sarah is now the same age Ginny was four years ago, Sarah is now eight.

Shakes all around (PAGE 14)

Each handshake counts as one, even though two people are shaking hands. So, when Doris shook hands with the other five children, that was five handshakes.

Then, child number two had his hand shaken by numbers 3, 4, 5, and 6, which is four more handshakes. Child 3, in addition to shaking hands with number 2 (which has already been

29

counted), shook hands with 4, 5, and 6, for three more handshakes. Number 4, in addition to shaking hands with 2 and 3, shook hands with 5 and 6, which is two more shakes. Numbers 5 and 6 had shaken hands with all the others, so when they shook hands with each other, that counts as the last shake.

That makes a total of fifteen handshakes.

Mr. Short, Mr. Tall, and Mr. Middle (PAGE 14)

Both the tall man and Mr. Short spoke to the lady. So, the tall man was not Mr. Short. And because his name doesn't match his appearance, he can't be Mr. Tall. So, he must be Mr. Middle.

If Mr. Middle is the tall man, Mr. Short can't be tall. And, of course, he can't be short. So, he has to be the middle-sized man. The name of the short man, then, has to be Mr. Tall.

Three puppies (PAGE 15)

The last two clues tell you that Ginger was in front of Spot, and Bingo was in front of Ginger. So, Bingo was first, followed by Ginger, then Spot.

Who earns the most? (PAGE 16)

It seems as if Bill, with his yearly $200 raise, should earn more. But Ellen will actually earn $50 a year more than Bill.

Bill gets his raise at the beginning of each year. So, he won't have a raise during his first year. He'll just make his $10,000 salary. But Ellen gets a $50 raise after her first six months. She earns $5,000 the first six months, and $5,050 the next six. That's a total of $10,050—which is $50 more than Bill will make.

At the start of his second year, Bill gets a $200 raise. He'll earn $10,200.

Ellen gets another $50 raise at the end of her second six months of work. So, her salary for the first six months of the second year will go up to $5,100. She gets another raise midway through the year, to $5,150. So, for her second year, she'll make $10,250 altogether. Again, that's $50 more than Bill will earn.

Ellen will always be $50 ahead. After five years, she'll have earned $250 more than Bill.

What's the number? (PAGE 17)

The car was traveling at a speed of 55 miles (or 55 kilometers) an hour. Thus, in two hours it will go 110 miles (or 110 kilometers). Add 110 to 11911 and you'll get 12021, the new backward-forward number that showed on the dial.

Squirrels and nuts (PAGE 18)

Inasmuch as the squirrels finished with 50 nuts each, you know, of course, that they started with a total of 100 nuts. To find out how many each squirrel had collected, you work backwards.

On the third try, Chitter gave Chatter 10 nuts, and they both had 50. That means that *before* the third try, Chitter must have had 60 nuts and Chatter had 40.

On the second try, Chitter gave Chatter as many nuts as Chatter already had. This left Chitter with 60 and Chatter with 40. So, before this, Chatter must have had 20 nuts and Chitter 80.

Chitter had the 80 nuts after the first try, when Chitter gave him as many nuts as Chatter already had. So Chitter must have started with 40 nuts, and Chatter gave him 40 more to make 80. And if Chitter started with 40 nuts, Chatter must have started with 60.

The coin test (PAGE 19)

The young girl must put three coins in each of the balance pans, setting the other three coins aside. If one pan sinks down, she will know that one of the coins in the other pan is a counterfeit. If the pans balance, the counterfeit must be one of the three coins she set aside.

When the girl knows which group of coins includes the counterfeit, she can then put two of the coins in this group in the balance, one in each pan. She sets the other coin aside. If one of the coins in the balance is heavier, the other one is the counterfeit. If they balance, the coin she put aside is the counterfeit.

Thus, using the balance only twice, the young girl can find out which coin is the counterfeit.

The five hobbyists (PAGE 20)

Clues 1, 2, and 4 tell us that the gardener cannot be Mrs. Kowalski, Ms. Schmidt, or Miss Fujimoto. Clue 5 reveals that Ms. Rodriguez isn't the gardener either. That leaves only Mrs. Robinson. She's the gardener.

Clue 3 tells us that neither Mrs. Kowalski nor Miss Fujimoto is the knitter. Clue 5 also shows that Ms. Rodriguez isn't the knitter. So, the only one left who can be the knitter is Ms. Schmidt.

From clue 5 we also know that Ms. Rodriguez isn't the pottery maker. Thus, she has to be either the painter or the writer. Clue 4 reveals that Miss Fujimoto isn't the writer, so she must be either the painter or pottery maker. As for Mrs. Kowalski, she might be the painter, the writer, or the pottery maker.

But clue 6 reveals that neither Mrs. Kowalski nor Ms. Rodriguez is the painter. Inasmuch as Ms. Rodriguez has to be either the painter or the writer, she is obviously the writer. With Mrs. Kowalski ruled out as both painter and writer, she must be the pottery maker. That leaves Miss Fujimoto as the painter.

So, the five hobbyists are:
- Mrs. Robinson, gardener
- Ms. Schmidt, knitter
- Ms. Rodriguez, writer
- Mrs. Kowalski, pottery maker
- Miss Fujimoto, painter.

Frank's folly (PAGE 22)

To solve this puzzle, work backwards.

At the end of the third block, Frank gave Larry eight pieces of candy and had none left. The eight pieces were what Frank had after his candy had been doubled. Eight is four doubled. So, at the end of the third block, Frank must have had four pieces.

$4 \times 2 = 8 - 8 = 0$

The four pieces were left after Frank gave *back* eight pieces at the end of the second block. Eight and four are twelve, so twelve pieces is what Frank had after Larry doubled Frank's candy. Twelve is six doubled. Thus, Frank had six pieces at the beginning of the second block.

$6 \times 2 = 12 - 8 = 4$

Those six pieces were left after Frank gave *back* eight pieces at the end of the first block. Eight and six are fourteen. Fourteen is what Frank had after his candy was doubled. Fourteen is seven doubled, so Frank started out with seven pieces of candy.

$7 \times 2 = 14 - 8 = 6$

The dull movie (PAGE 23)

At the beginning of the movie, there were 150 people in the audience.

To get this answer, you have to work backwards from the number of people, 25, that were still in the theater when the movie ended.

You know that there were 25 people left after *half* the people walked out during the last part of the movie. As 25 is half of 50, there must have been 50 people in the theater when the last part of the movie began.

You know that half the people in the audience left during the middle part of the movie. Thus, the 50 people still there were the other half. So, there must have been 100 people when the middle part of the movie began.

You know that one-third of the whole audience left during the first part of the movie. Thus, the 100 people who sat through the middle part of the movie were *two-thirds* of the whole audience. If you divide 100 by 2, you will get 50, which is one-third of the *total* audience. Therefore, 50 people left during the first part of the movie, and 50 plus 100 is 150—the number of people in the audience when the movie began.

A nutty puzzle (PAGE 24)

To solve this puzzle, work backwards.

After making all his sales, Nicodemus had one dozen, or 12 nuts left. His third customer, the little girl, had bought one-third of what he had before that, so the twelve nuts were the other two-thirds. One-third is obviously six nuts. Thus, Nicodemus had 18 nuts (6 x 3 = 18) before making the last sale.

Those 18 nuts were left after the second customer bought half of Nicodemus's nuts plus half a nut. Therefore, the 18 nuts must be half, *less one-half a nut,* of what Nicodemus had before his second sale. So, Nicodemus must have had 37 nuts—half of 37 is 18 ½, plus a half is 19. The second customer bought 19 nuts, leaving Nicodemus with 18.

The 37 nuts were left from the first sale. The first customer bought half of Nicodemus's nuts, plus half a nut. So, the 37 nuts must be half, less half a nut, of what Nicodemus started with. Nicodemus must have started with 75 nuts. And the first customer bought 38—half of 75 is 37 ½, plus a half is 38.

The lunch bunch (PAGE 25)

The first clue tells us Mark had the same things Mary Jo did, except for an ice-cream sundae. So, we know that one of the things Mary Jo had was an ice-cream sundae.

Clue 2 tells us Diana had what Brian did, plus a hamburger. So, we know that Diana had a hamburger.

Clue 3 says that Gail had what Judy had, but without French fries. Obviously, Judy had French fries as part of her lunch.

Clue 4 tells us Mary Jo had what Judy had, but with pop instead of a milk shake. Thus, we now know that Mary Jo had French fries as Judy did, plus pop. We also found that Judy had a milk shake with her fries. And as clue 3 told us that Gail had what Judy did, but without fries, Gail had a milk shake, too.

Clue 5 says Brian had what Matt did, except that he did not have a hamburger. This tells us Matt had a hamburger. The very first clue told us Matt had the same things as Mary Jo, except for a sundae, so we can see that Matt had French fries and pop with his hamburger. And since he and Mary Jo had the same things, she must have had a hamburger, too. As for Brian, if he had what Matt did, without the hamburger, then he had French fries and pop. And if we go back to the second clue, we see that Diana had what Brian did—French fries and pop—plus a hamburger.

So, what the Lunch Bunch had was:
- Mary Jo: pop, hamburger, French fries, sundae
- Diana: pop, French fries, hamburger
- Judy: milk shake, French fries
- Gail: milk shake
- Matt: pop, French fries, hamburger
- Brian: pop, French fries

What's missing? (PAGE 26)

The missing things are shown in blue.

A hungry bookworm (PAGE 28)

When two books are side by side on a shelf, the *front* cover of the first book, the book on the left, is pressed against the *back* cover of the second book. A book's first page is next to its front cover, and its last page is next to its back cover. So, if the bookworm ate through the first page of the first book, it then ate its way through the book's front cover. Next, it ate its way through the back cover of the second book. Then it ate through the last page of the second book. So, altogether, it ate its way through only two pages.